Equations

It balances 1

A1

The apples weigh 2 kg.

A2 The calf weighs 70 kg.

A3

These scales will balance if you add 2 kg on the right-hand side.

A4 The fourth weight should be 5 kg.

A5 (a) The rabbit weighs 4 kg.
(b) The hedgehog weighs 5 kg – a giant hedgehog!
(c) The baby tiger weighs 25 kg. (How would you weigh a wild animal like a tiger?)

A6 Each sack weighs 11 kg. (Could you easily lift a sack weighing this?)

A7 Both boxes weigh 28 kg, so one box weighs 28 ÷ 2 = 14 kg.

A8 Carla could weigh all these boxes:
1, 2, 3, 4, 5, 6 and 7 kg.

A9 If she has 1, 3 and 6 kg weights only, she can balance these weights of box.

Left-hand side	Right-hand side
a 1 kg box	1 kg
a 2 kg box + 1 kg	3 kg
a 3 kg box	3 kg
a 4 kg box	3 kg + 1 kg
a 5 kg box + 1 kg	6 kg
a 6 kg box	6 kg
a 7 kg box	6 kg + 1 kg
a 8 kg box + 1 kg	6 kg + 3 kg
a 9 kg box	6 kg + 3 kg
a 10 kg box	6 kg + 3 kg + 1 kg

A10 The total weight of the right-hand side is 21 + 13 + 20 + 21 = 75 kg.
Five bottles weigh 75 kg, so one bottle weighs 75 ÷ 5 = 15 kg.

A11 Check these very carefully. See your teacher if you cannot work out where you are going wrong.
(a) Yes, it will still balance if you add 1 kg to each side.
(b) Yes, it will still balance if you add 2 kg to each side.
(c) Yes, it will still balance if you take 2 kg away from each side.
(d) It will not balance it you take 1 kg from one side and put it on the other side. (You have treated the sides differently.)

A12 (a) Yes, swapping the sides will give a balanced set of scales.
(b) Yes, halving the amount on each side will give a balanced set of scales.
(c) Yes, doubling the amount on each side will give a balanced set of scales.

B1

The cat weighs 5 kg.

B2 The rabbit weighs 4 kg.

B3

(a) The cat weighs 4 kg.
(b) If you take 2 kg from each side, then 3 kg will balance with the penguin.
(c) Take 5 kg from each side. The puppy weighs 4 kg.
(d) Take 1 kg from each side. The hedge-hog weighs 4 kg.

B4 (a) Taking a jar from each side will not make any difference – the scales will still balance.
(b) A jar weighs 5kg.

B5 If you take a bottle from each side, then 3 kg will balance with one bottle – one bottle weighs 3 kg.

B6 (a) ▲

Take two tins from each side.
One tin weighs 2 kg.
(b) Take two tins from each side.
One tin weighs 5 kg.
(c) Taking two tins from each side will balance two tins with 6 kg. So one tin weighs 3 kg.
(d) If 2 kg is removed from each side, three tins balance with 6 kg. So one tin weighs 2 kg.

C1 Each tin weighs 4 kg. (Take a tin from each side and then 3 kg from each side.)

C2

(a) A tin weighs 2 kg.
(b) The left side weighs
2 kg (one tin) + 6 kg = 8 kg.
(c) The right side weighs
6 kg (three tins) + 2 kg = 8 kg.
Making sure that both sides weigh the same is a good way of checking your answer for the weight of one tin.

C3 (a) One tin weighs 10 kg.
(b) Left side weighs
20 kg (two tins) + 11 kg = 31 kg.
Right side weighs
30 kg (three tins) + 1 kg = 31 kg.
Both sides are the same!

C4 (a) Take one tin from each side.
12 kg balances three tins, so one tin weighs 4 kg.

(b) Check:
Left side of balance:
4 kg (one tin) + 12 kg = 16 kg.
Right side of balance:
4 tins, each weighing 4 kg = 16 kg.
Both sides weigh the same.

C5 (a) Take a tin and 3 kg from each side to give a tin balancing 1 kg, so one tin weighs 1 kg.
▲
(b) Left side of balance:
2 kg (two tins) + 3 kg = 5 kg
Right side of balance:
1 kg (one tin) + 4 kg = 5 kg.
Both sides weigh the same.

D1 (a) ? + 4 = 7
so ? = 3
(b) ? + ? + ? + ? + ? + 5 = ? + ? + ? + 11
Take three ?s from each side.
? + ? + 5 = 11
Take 5 kg from each side.
? + ? = 6
so ? = 3
(c) ? + ? + ? + ? + ? + 5 = ? + ? + 17
Take two ?s from each side.
? + ? + ? + 5 = 17
Take 5 kg from each side.
? + ? + ? = 12
so ? = 4

D2 You should already have checked this with the working on the inside back cover of *It balances 1*.

D3

? + ? + 1 = ? + 4
Take a ? from each side,
which is the same as
? + 1 = 4.
Take 1 from each side,
which is the same as
? = 3.

D4 $? + 6 = ? + ? + 3$

Take a $?$ from each side, which is the same as

  $6 = ? + 3.$

Take 3 from each side, which is the same as

  $3 = ?.$

Which is the same as

  $? = 3.$

(Imagine looking at a balance from the other side.)

D5

(a) $?$ stands for the weight of a bottle.

  $? + ? + ? + 5 = ? + ? + ? + ? + 3$

      $5 = ? + 3$

      $2 = ?$

    so $? = 2$

(b) $?$ stands for the weight of one suit-case.

  $? + ? + ? + 17 = ? + ? + ? + ? + ? + 5$

      $17 = ? + ? + 5$

      $12 = ? + ?$

      $6 = ?$

    so $? = 6$

E1 Check: left-hand side

      $? + ? + 5 = 3 + 3 + 5 = 11$

    right-hand side

      $? + ? + ? + 2 = 3 + 3 + 3 + 2 = 11$

E2 $? + 14 = ? + ? + 3$

      $14 = ? + 3$

      $11 = ?$

    so $? = 11$

Check: left-hand side

      $? + 14 = 11 + 14 = 25$

    right-hand side

      $? + ? + 3 = 11 + 11 + 3 = 25$

E3 $* + * + 2 = * + 7$

      $* + 2 = 7$

      $* = 5$

Check: left-hand side

      $* + * + 2 = 5 + 5 + 2 = 12$

    right-hand side

      $* + 7 = 5 + 7 = 12$

E4 (a) $? + ? + ? + 2 = ? + ? + 3$

        $? + 2 = 3$

        $? = 1$

(b) $\blacktriangle + \blacktriangle + \blacktriangle + \blacktriangle = \blacktriangle + \blacktriangle + 10$

        $\blacktriangle + \blacktriangle = 10$

        $\blacktriangle = 5$

(c) $\blacksquare + 5 + \blacksquare = 11 + \blacksquare$

        $\blacksquare + 5 = 11$

        $\blacksquare = 6$

(d) $\blacktriangledown + \blacktriangledown + \blacktriangledown + 7 = \blacktriangledown + \blacktriangledown + 15$

        $\blacktriangledown + 7 = 15$

        $\blacktriangledown = 8$

(e) $6 + x + x + x = x + 32$

      $6 + x + x = 32$

        $x + x = 26$

        $x = 13$

F1

$?$ stands the number of people in a minibus.

Men in one minibus plus 22 walkers.

So, $? + ? + ? = ? + 22$

Number of women in the three minibuses.

  $? + ? + ? = ? + 22$

    $? + ? = 22$

      $? = 11$

So a minibus holds 11 people.

F2

$?$ stands for the number of wine gums in a tube.

Jane has $? + ? + ? + 1$ wine gums.

Peter has $? + 19$ wine gums.
They both have the same number
of wine gums.
So, $? + ? + ? + 1 = ? + 19$ (Don't worry if
your wrote $? + 19 = ? + ? + ? + 1$)

$$? + ? + 1 = 19$$
$$? + ? = 18$$
$$? = 9$$

there are 9 wine gums in a tube.

F3 $?$ stands for the number of biscuits in a
packet.
Winnie has $? + ? + ? + ? + 2$
Sam has $? + ? + 16$
They both have the same number of
biscuits.
So, $? + ? + ? + ? + 2 = ? + ? + 16$
(Don't worry if you wrote
$? + ? + 16 = ? + ? + ? + ? + 2$)

$$? + ? + 2 = 16$$
$$? + ? = 14$$
$$? = 7$$

there are 7 biscuits in a packet.

F4 $?$ stands for the number of pencils in a box.
Christine has $? + ? + ? + 2$ pencils.
Sonya has $? + 12$ pencils.
They both have the same number of pencils.
So we can write this equation as:

$$? + ? + ? + 2 = ? + 12$$

The solution to this equation is $? = 5$,
so a box contains 5 pencils.

F5   Let $?$ stand for the number of people in a
single coach.
(You might have used another symbol.)
$? + ? + ? + 4$ people go to the concert.
$? + 50$ people leave the concert.
The same number left as went to the
concert (no one got lost!).
So $? + ? + ? + 4 = ? + 50$
The solution to this equation is $? = 23$.
Each coach holds 23 passengers.

F6   Use X to stand for the number of
chocolates in a box.
When Peter started there were
$X + X + X + X + 2$ chocolates.
When Joan finished there were
$X + X + 44$ chocolates.
If no one has eaten any chocolates there

should be as many for Joan as for Peter!
We can write this equation

$$X + X + X + X + 2 = X + X + 44$$

This equation has the solution $X = 21$.
So, each box holds 21 chocolates.

F7   Your own stories.

Challenges

You may have used different symbols in
your equations.

(1) If z stands for the number of people
     in a boat,
     $$z + z + z + 4 = z + z + 21$$
     The solution is $z = 17$, each boat
     holds 17 people.

(2) If each barrel holds x litres of water
     $$x + 32 = x + x + x + 6$$
     This has the solution $x = 13$.
     A barrel holds 13 litres of water.

(3) If there are $?$ apples in a box,
     $$? + ? + ? + ? + 5 = ? + 26$$
     The solution to this equation, $? = 7$,
     shows that each box holds 7 apples.

It balances 1: extension

A1

```
? + ? + ? + 5 = ? + 11
    ? + ? + 5 = 11
        ? + ? = 6
            ? = 3
```

So
$? = 3$

A2   (a) $? + ? + ? + ? + 1 = ? + 7$
                $? + ? + ? + 1 = 7$
                    $? + ? + ? = 6$
                        so $? = 2$

(b) Check: left-hand side
     $? + ? + ? + ? + 1 = 2 + 2 + 2 + 2 + 1 = 9$
     right-hand side
     $? + 7 = 2 + 7 = 9$

A3 (a) $? + ? + ? + 2 = ? + 10$
$? + ? + 2 = 10$
$? + ? = 8$
$? = 4$

(b) $? + ? + 73 = ? + 95$
$? + 73 = 95$
$? = 95 - 73 = 22$

(c) $? + ? + ? + ? + ? + 6 = ? + 42$
$? + ? + ? + ? + 6 = 42$
$? + ? + ? + ? = 36$
$? = 9$

(d) You should already have checked this.
(Inside the back cover of the booklet.)

(e) $? = 3$

(f) $? + 12 = ? + ? + ? + ? + 3$
$12 = ? + ? + ? + 3$
$9 = ? + ? + ?$
$3 = ?$
$? = 3$

(g) $? + ? = 9$, $? = 4\frac{1}{2}$

(h) $35 = ? + ?$, $? = 17\frac{1}{2}$

B1 When $3(?) = 66$, then $? = 22$
(Check: $3(22) = 66$)

B2

(b) $4(?) = 28$ (c) $5(?) = 135$ (d) $3(?) = 168$
$? = 7$ $? = 27$ $? = 56$

B3 (a) $6(?) = 18$ (b) $10(?) = 80$
$? = 3$ $? = 8$
(c) $39 = 3(?)$ (d) $5(?) = 65$
$13 = ?$, so $? = 13$ $? = 13$
(e) $64 = 4(?)$ (f) $2(?) = 132$
$16 = ?$, so $? = 16$ $? = 66$
(g) $7(?) = 91$ (h) $9(?) = 117$
$? = 13$ $? = 13$
(i) $264 = 12(?)$
$22 = ?$, so $? = 22$

B4

Don't worry if you've used different symbols.

(a) Let X stand for the weight of one tin.
$4(X) = 2(X) + 8$
$2(X) = 8$
so $X = 4$

Check:
Left-hand side $= 4(4) = 16$
Right-hand side $= 2(4) + 8 = 8 + 8 = 16$

(b) Let ? stand for the weight of one box.
$6(?) = 2(?) + 24$
$4(?) = 24$
so $? = 6$

Check:
Left-hand side $= 6(6) = 36$
Right-hand side $= 2(6) + 24 = 12 + 24 = 36$

B5 (a) $10(\blacktriangle) + 1 = 4(\blacktriangle) + 19$
$6(\blacktriangle) + 1 = 19$
$6(\blacktriangle) = 18$
$\blacktriangle = 3$

(b) $7(\blacksquare) + 6 = 2(\blacksquare) + 41$
$5(\blacksquare) + 6 = 41$
$5(\blacksquare) = 35$
$\blacksquare = 7$

(c) $2(?) + 23 = 5(?) + 5$
$23 = 3(?) + 5$
$18 = 3(?)$
$3(?) = 18$
$? = 6$

(d) $11(\blacksquare) + 4 = 4(\blacksquare) + 25$
$7(\blacksquare) + 4 = 25$
$7(\blacksquare) = 21$
$\blacksquare = 3$

(e) $13(\blacktriangleright) + 11 = 15(\blacktriangleright) + 3$
$11 = 2(\blacktriangleright) + 3$
$8 = 2(\blacktriangleright)$
$\blacktriangleright = 4$

(f) $3(X) + 4 = X + 16$
$2(X) + 4 = 16$
$2(X) = 12$
$X = 6$

(g) $\blacksquare + 12 = 3(\blacksquare)$
$12 = 2(\blacksquare)$
$\blacksquare = 6$

(h) $9(?) + 17 = 8(?) + 28$
$? + 17 = 28$
$? = 11$

5

(i) $36 + 8(x) = 11(x)$
$$36 = 3(x)$$
$$3(x) = 36$$
$$x = 12$$
(j) $5(X) + 23 = 7(X) + 18$
$$23 = 2(X) + 18$$
$$5 = 2(X)$$
$$2(X) = 5$$
$$X = 2\tfrac{1}{2}$$

C1 The equation you need to solve is:
$$X + X + X = 30 + X$$
$$3(X) = 30 + X$$
$$2(X) = 30$$
$$\text{so } X = 15$$

C2 (a) X stands for the number of felt tips in a packet. Mark has $X + 12$ pens and Sanjay $3(X) + 2$. They both have the same number of pens, so
$$X + 12 = 3(X) + 2$$
(This equation needs to be solved to find X.)
$$X + 10 = 3(X)$$
$$10 = 2(X)$$
$$\text{so } X = 5$$
(b) We can work out how many pens each person has by using either $X + 12$, or $3(X) + 2$. They must both give the same number!
When $X = 5$, $X + 12 = 17$ and
$3(X) + 2 = 15 + 2 = 17$.
Each person has 17 felt tips.

C3 Let P stand for the number of sheep in a pen (you may have used another symbol). The equation to be solved is
$$5(P) + 3 = 3(P) + 15$$
$$5(P) = 3(P) + 12$$
$$2(P) = 12$$
$$\text{so } P = 6.$$
There were six sheep in a pen.

C4 If each barrel holds X litres of water then
$$X + 37 = 5(X) + 4$$
$$X + 33 = 5(X)$$
$$33 = 4(X)$$
$$\text{so } 4(X) = 33$$
$$\text{so } X = 8 \cdot 25$$
$(33 \div 4)$, a barrel holds $8 \cdot 25$ litres.

D1 (a) Let X stand for the number of horses in the field at first.
(b) There are now $X + 33$ horses.
(c) 4 times as many as before, $4X$.
(d) So $X + 33 = 4X$
$$\text{So } 33 = 3X$$
$$\text{So } 11 = X$$
So there were 11 horses in the field at first.
Check: when $X = 11$, $X + 33 = 44$ and $4(X) = 44$

D2 (a) Let X be the number of cars in the car park at first.
(b) 54 more arrived, so there were now $X + 54$ cars.
(c) 10 times the number that were there first is $10(X)$.
(d) So $X + 54 = 10(X)$.
This equation has the solution $X = 6$. There were 6 cars in the car park at first. Check: When $X = 6$, $X + 54 = 60$ and $10(X) = 60$

D3 Let ? stand for the number of giraffes at first. The story gives this equation $? + 40 = 6(?)$ This has a solution $? = 8$. So there were 8 giraffes at the water hole at first.

D4 Let X stand for the number of passengers on the ship at first. When 152 more get on there were $X + 152$ passengers on board. This was 5 times the number at first, which is $5(X)$.
If no one fell overboard $X + 152 = 5(X)$. Solving this equation gives $X = 38$. There were 38 passengers on board at first.

D5 Your own problem.

D6 Your own equations which have 4 as a solution.

E1 (a) The bottom output is $3(?) + 40$.
(b) The solution to the equation
$5(?) + 6 = 3(?) + 40$ is $? = 17$.
So the input number was 17.
Check: When $? = 17$
$5(?) + 6 = 85 + 6 = 91$ and
$3(?) + 40 = 51 + 40 = 91$.

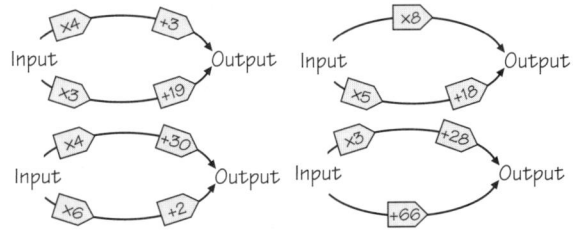

Let ? be the input number for each pair of number machines.

(a) The equation to be solved is
$4(?) + 3 = 3(?) + 19$. The solution is
$? = 16$, the input number.

(b) The input number is the solution to
$8(?) = 5(?) + 18$. This is $? = 6$.

(c) You need to solve the equation
$4(?) + 30 = 6(?) + 2$. The solution (the input number) is $? = 14$.

(d) For the top machine $3(?) + 28$ and for the bottom one $? + 66$. They both give the same output, so $3(?) + 28 = ? + 66$. Solving this equation gives $? = 19$.

The top machine gives $3 \times 3 + 6$, which is 15.
The bottom machine chain must give an output of 15.
So $3(\blacktriangle) + \blacksquare = 15$.
It's probably easiest to use a table.
Here are just a few values of $\blacktriangle$ and $\blacksquare$ which fit the equation

$\blacktriangle$	$3(\blacktriangle)$	$\blacksquare$	$3(\blacktriangle) + \blacksquare = 15$
1	3	12	$3(1) + 12 = 15$
2	6	9	$3(2) + 9 = 15$
3	9	6	$3(3) + 6 = 15$
4	12	3	$3(4) + 3 = 15$
5	15	0	$5(3) + 0 = 15$
6	18	⁻3	$6(3) + ⁻3 = 15$
7	21	⁻6	$7(3) + ⁻6 = 15$

There are many many more.
What about fractions or decimals?

 Challenges

(1)

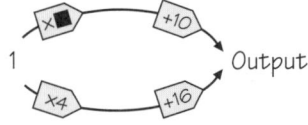

We can find an expression for the top chain ($\blacksquare + 10$).
This must be equal to the bottom chain. ($1 \times 4 + 16 = 20$)
So $\blacksquare + 10 = 20$, this means that $\blacksquare$ is equal to 10.

(2)

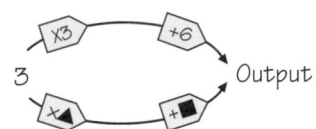

It balances 2

A1

? stands for the weight of one can.
(a) $? + ? + ? + 6 = ? + ? + ? + ? + 3$
(b) Taking three ?s from each side gives
$6 = ? + 3$
Taking 3 from each side gives
$3 = ?$
So $? = 3$, each can weighs 3 kg.

Check your own answers to the worksheet (if you did it).

A2 Here is the complete solution to the equation.
$? + ? + ? + ? + 12 = ? + ? + ? + 20$
Take 3 ?s from each side
so $? + 12 = 20$
Take 12 from each side
so $? = 8$

A3 (a) Let ? stand for the
weight of one cake.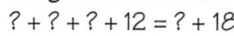
$? + ? + ? + 12 = ? + 18$
Take a ? from each side.
$? + ? + 12 = 18$
Take 12 from each side.
$? + ? = 6$
so $? = 3$

(b) Let ? stand for the
weight of each case.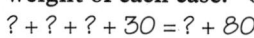
$? + ? + ? + 30 = ? + 80$
Take a ? from each side.
$? + ? + 30 = 80$
Take 30 from each side.
$? + ? = 50$
so $? = 25$

(Could you easily carry a case this heavy?)

A4 (a) $? + ? + ? + 5 = ? + ? + 6$
Take 2 ?s from each side.
so $? + 5 = 6$
Take 5 from each side.
so $? = 1$

(b) $? + ? + ? + ? = ? + ? + 24$
Take 2 ?s from each side.
$? + ? = 24$
so $? = 12$

(c) $* + 10 + * = * + 12$
*Take * from each side.*
$* + 10 = 12$
Take 10 from each side.
so $* = 2$

(d) $9 + \blacklozenge + \blacklozenge + \blacklozenge = 15 + \blacklozenge + \blacklozenge$
Take two $\blacklozenge$ from each side.
$9 + \blacklozenge = 15$
Take 9 from each side.
so $\blacklozenge = 6$

(e) $x + x + 6 = x + x + x + x$
Take 2 xs from each side.
$6 = x + x$
so $x = 3$

think carefully about this one

(f) $8 + ? + ? + 3 = 2 + ? + 15$
Take a ? from each side.
$8 + ? + 3 = 2 + 15$
Tidy up the numbers.
$? + 11 = 17$
Take 11 from each side.
so $? = 6$

B1 ▲ When we group symbols together it is
sometimes called simplifying.
(a) $? + ? + ? + ? = 24$ can be written $4(?) = 24$
(b) $? + ? + ? + ? + ? = 20$ is $5(?) = 20$
(c) $? + ? = ? + 30$ is $2(?) = ? + 30$
(d) $? + ? + 10 = 62$ is $2(?) + 10 = 62$
(e) $48 + ? = ? + ? + ?$ is $48 + ? = 3(?)$

B2 (a) 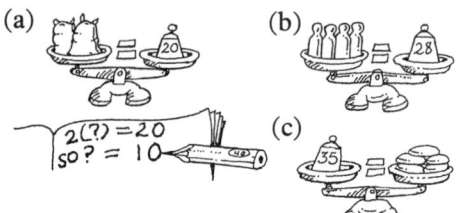 (b)

(c)

(a) This has already been done for you.
(b) ? stands for the weight of one bottle
in kilograms.
$4(?) = 28$
so $? = 7$
(c) ? stands for the weight of one bun
in kilograms. (They are very heavy
buns!)
$35 = 5(?)$
so $7 = ?$
so $? = 7$ (They must be 15 pounders!)

B3 (a) $6(?) = 18$ (b) $10(?) = 80$ (c) $15 = 3(?)$
$? = 3$ $? = 8$ $5 = ?$
so $? = 5$
(d) $5(?) = 45$ (e) $24 = 6(?)$ (f) $2(?) = 100$
$? = 9$ $4 = ?$ $? = 50$
so $? = 4$

B4 (a) $5(?) + 6 = 7(?)$ (b) $7(?) = 4(?) + 6$
so $6 = 2(?)$ *so* $3(?) = 6$
so $? = 3$ *so* $? = 2$
(c) $15 = 5(?)$ (d) $12 = 2(?)$
so $3 = ?$ *so* $6 = ?$
so $? = 3$ *so* $? = 6$
(e) $24 = 6(?)$ (f) $10(?) = 30$
so $? = 4$ *so* $? = 3$

B5 ▲ (a) $3(?) = ? + 8$ (b) $? + 7 = 2(?)$
$2(?) = 8$ $7 = ?$
so $? = 4$ *so* $? = 7$

$\mathcal{N}u^m\mathsf{ber}$ Puzzles

- $3(?) + 4 = 19$
  $3(?) = 15$
  so $? = 5$.
  The number was 5.
  Check to see that it fits the puzzle.

- Here is the equation which fits.
  "Twice the number added to three times the number gives 55."
  Remember ? stands for the number.
  $2(?) + 3(?) = 55$
  Here is how to find the solution to the equation:
  $5(?) = 55$ ($2(?) + 3(?)$ gives $5(?)$)
  so $? = 11$.
  This is the number.

- "Three times a certain number is three less than 30."
  If ? stands for the number, then we can write the equation:
  $3(?) = 27$ (27 is three less than 30)
  so $? = 9$

B6 (a), (b) and (c)
  $5(?) + 3 = 2(?) + 9$
  so $3(?) + 3 = $ $\quad$ 9 (Brian took $2(?)$ from each side.)
  so $3(?)$ $\quad = \quad$ 6 (He took 3 from each side.)
  $\qquad$ so $? = 2$

B7 (a) $3(?) + 7 = 5(?) + 3$
  Take $3(?)$ from each side.
  $\quad 7 = 2(?) + 3$
  Take 3 from each side.
  $\quad 4 = 2(?)$
  $\quad$ so $? = 2$
  Check: left-hand side:
  $3(?) + 7 = 3(2) + 7 = 13$
  right-hand side:
  $5(?) + 3 = 5(2) + 3 = 13$
  (b) $7(?) + 5 = 3(?) + 9$
  Take $3(?)$ from each side.
  $\quad 4(?) + 5 = 9$
  Take 5 from each side.
  $\quad 4(?) = 4$
  $\quad$ so $? = 1$

Check: when ? stands for 1
  left-hand side $7(?) + 5 = 7(1) + 5 = 12$
  right-hand side $3(?) + 9 = 3(1) + 9 = 12$

If you get any of the rest wrong, check through your working.

If you are still not sure where you are going wrong, then ask your teacher.
  (c) $3(?) + 10 = 4(?) + 2$, the solution is $? = 8$.
  (d) $12 + 3(?) = 4 + 5(?)$, the solution is $? = 4$.
  (e) The solution to the equation
     $25 + 6(?) = 3(?) + 43$ is $? = 6$.
  (f) $12(?) + 42 = 4(?) + 138$
     has the solution $? = 12$.

C1

? stands for the weight of an animal.
  (a) $2(?) + 5 = 9$
     $\quad 2(?) = 4$
     $\quad$ so $? = 2$
  (b) $\quad 11 = 2(?) + 5$
     $\quad 6 = 2(?)$
     $\quad$ so $? = 3$

C2 (a) $3(?) + 1 = 10$
▲ $\qquad 3(?) = 9$
  $\qquad$ so $? = 3$
  (b) $2 + 2 + 2 + 2 + 2 = ? + 1 + 2(?)$
     Tidy up the numbers and ?s.
     (You may have done this differently.
     Don't worry if you have.)
     $\quad 10 = 3(?) + 1$
     Take 1 from each side
     $\quad 9 = 3(?)$
     $\quad$ so $? = 3$

(c) $X + X + X + 1 = 7 + 3$
Tidy up
$$3(X) + 1 = 10$$
Take 1 from each side.
$$3(X) = 9$$
$$so \; X = 3$$

(d) $\blacksquare + \blacksquare + \blacksquare + \blacksquare + \blacksquare = 4(\blacksquare) + 20$
Take 4 $\blacksquare$s from each side.
$$so \; \blacksquare = 20$$

C3 Your own equations or puzzles which have 2 as a solution.

It balances 3

A1 See page 2 of the booklet, you have already marked it!

A2
▲

If n stands for the number of mints in a packet then Stephen has $5n + 2$ mints and Susan has $3n + 20$. (Stephen has 5 whole packets and Susan 3.)
Stephen and Susan have the same number of mints, so
$$5n + 2 = 3n + 20$$
(we need to solve this equation.)
$$2n + 2 = 20$$
$$2n = 18$$
$$so \; n = 9$$
this tells us that there are 9 mints in a packet.

A3 (a) Let n stand for the number of sheets in a pad.
Sandra has 4 pads and 78 extra sheets. This makes a total of $4n + 78$ sheets of paper. Nancy has 9 pads and 3 extra sheets so she has $9n + 3$ sheets of paper in all. The girls have the same total number of sheets.

This means we can write
$$4n + 78 = 9n + 3$$
$$so \; 78 = 5n + 3$$
$$so \; 75 = 5n$$
$$so \; n = 15 \qquad (75 \div 5 = 15)$$
There are 15 sheets of paper in each pad.

(b) Sandra has of a total of
$4 \times 15 + 78 = 60 + 78 = 138$ sheets, and Nancy has
$9 \times 15 + 3 = 135 + 3 = 138$ sheets.
If we had found that the girls had a different number of sheets it would have shown that we had made a mistake!

A4

(a) $7L + 22 = 3L + 226$
Take $3L$ from each side.
$$4L + 22 = 226$$
Take 22 from each side.
$$4L = 204$$
$$so \; L = 51$$
You should check your answer by testing that the left-hand side of the equation equals the right-hand side when L is 51.

(b) The planks were of equal length. When 51 is put in place of L in the expression for the length of one plank $7L + 22$ it comes to 379. Using the other expression, $3L + 226$, we also get 379.

A5

L stands for the length of a jail window bar.

Ms Bone cuts 4 bars and has 60 cm left, so the total length of the one rod is $4L + 60$ cm. From the second (equal length) rod she cuts 2 bars and has 230 cm left.

We can write this total length as $2L + 230$.

Both lengths of iron were the same so $4L + 60 = 2L + 230$.

$$2L + 60 = 230$$
$$2L = 170$$
$$\text{so } L = 85$$

This shows that each of the bars is 85 cm long.

A6 Let each van hold *n* people. (It does not matter if you used a different symbol.)

Going to the concert there were $4 + 6 + 3n$ people (or $10 + 3n$). (4 went by car, 6 walked, the rest filled up 3 vans.)

$6 + 9 + 2n$ (or $15 + 2n$) returned. (6 by car, 9 walked, the rest filled up 2 vans.)

If no one was lost, the number going must be equal to the number returning.

$$10 + 3n = 15 + 2n$$
$$10 + n = 15$$
$$n = 5 \qquad \text{Each van holds 5 people.}$$

A7

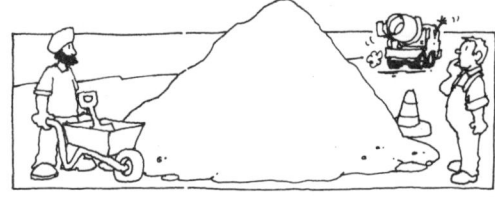

Let each wheelbarrow hold *W* bucketfuls of cement. (You might have used a different symbol.)

The Singhs have $12W + 10$ bucketfuls. The Rosses have $9W + 37$ bucketfuls. Both families have equal shares of the cement. So we can write this equation $12W + 10 = 9W + 37$.

We need to solve this equation to find *W*, how many bucketfuls a wheelbarrow holds.

$$12W + 10 = 9W + 37$$
$$3W + 10 = 37$$
$$3W = 27$$
$$\text{so } W = 9$$

B1

Julie had some records.

Her friend gave her 252 more.

Julie now had 5 times as many records as before.

(a) Let *n* stand for the number of records Julie had to start with.

(b) She now has $n + 252$ records.

(c) She now has $5n$ records.

(d) So $n + 252 = 5n$
$$252 = 4n$$
$$63 = n$$
$$n = 63$$

So Julie started off with 63 records.

Make sure you checked your answer!

B2
▲

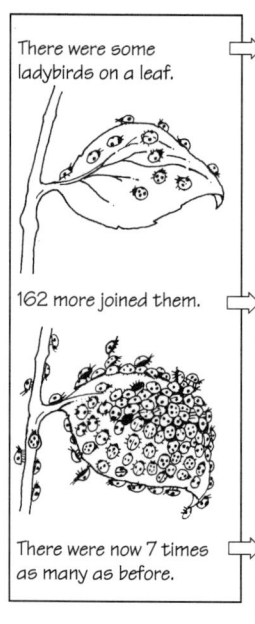

There were some ladybirds on a leaf.

162 more joined them.

There were now 7 times as many as before.

(a) Let n stand for the number of lady-birds on the leaf at the start.

(b) There are now $n + 162$ ladybirds.

(c) There are now $7n$ ladybirds.

(d) So $n + 162 = 7n$
$$162 = 6n$$
$$27 = n$$
$$n = 27$$

so there were 27 ladybirds on the leaf to start with.

Getting closer all the time

Your own methods.

A1 Your own results.

A2

Input	Output
15	13 *too big*
5	8 *too small*

Looking at the table, 15 is too big and 5 too small. A good trial would be about 11 which is a little closer to 15 than to 5.

A3

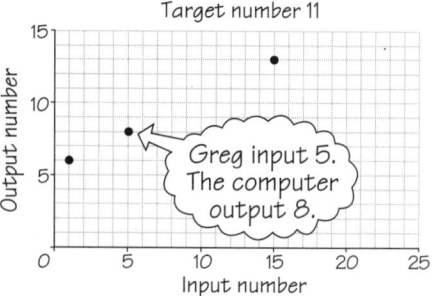

Target number 11

Greg input 5. The computer output 8.

If you look at the points already on Greg's graph they seem to make a straight line. So to get an output of 11, you would probably need to input 11.

A4 It's your choice!

A5 Not all points give a straight line when they are joined up.

B1 It probably helps to use a table for
▲ questions like these.

(a) *Five times the number added to twice the number gives 84.*

Number	5 times	Twice	5 times added to twice
6	30	12	42 (*too small*)
10	50	20	70 (*too small*)
15	75	30	105 (*too big*)
12	60	24	**84** (*that's the one!*)

(You may have found the number quicker than this.)

(b) *A number multiplied by itself (or squared) added to five gives 30.*

Number	Number squared	Number squared added to 5
3	9	14 (*too small*)
5	25	**30** (*that's the one!*)

(c) *Twice a number added to half the number gives 25.*

Number	Twice the number	Half the number	Twice added to half
6	12	3	15 (*too small*)
14	28	7	35 (*too big*)
10	20	5	25 (*that's the one!*)

B2 Your own riddles.

B3 Here is one way, you may well have used a different method yourselves.

A cubit is about 50 cm.

The problem is about land, so measuring to the nearest tenth (0·1) of a cubit is accurate enough.

(In fact an answer to the nearest whole cubit (50 cm) would probably do.)

You need to find by trial and improvement a number which when it is squared gives an answer as close as possible to 200.

Side of square (in cubits)	Area of square (in square cubits)
10	100
20	400
15	225
14	196
14·5	210·25
14·1	198·81
14·2	201·64

So the servant's square field has sides between 14·1 and 14·2 cubits. (14·17 would be a closer answer, but it would probably not be worthwhile to measure a field as accurately as this.)

B4 *"A number and a seventh of it added together make 19. What is the number?"*

Here is the way to find the number.

Number	Seventh of number (number ÷ 7)	Number + seventh of it
14	2	16
21	3	24
(*it looks as if we shall need decimals for this*)		
14·7	2·1	16·8
16	2·285 714 3	18·285 714
17	2·428 571 4	19·428 571
16·8	2·4	19·2 (*getting close now!*)
16·5	2·357 142 9	18·857 143
16·6	2·371 428 6	18·971 429
16·65	2·378 571 4	19·028 571

The number 16·65 gives a result of 19·028 571 which is close enough. Did you manage to get closer?

Was it a good idea to keep all the numbers after the decimal place?

If you know about loops in BASIC you could have written a short program to solve the riddle in B4

Here is a short program which would have done the job. See if you can figure out how it works. Try it if you have time.

```
10 FOR NUMBER = 16 TO 17 STEP 0·01
20 PRINT NUMBER, NUMBER + NUMBER/7
30 NEXT NUMBER
```

C1 What you are looking for is a number which gives 2·44 when it is squared (or multiplied by itself) and 1 added to the result. Here is one way.

Number	Number squared	Number squared + 1
1	1	2
2	4	5
1·1	1·21	2·21
1·2	**1·44**	**2·44**

C2 You should have found that when you entered 30 and pressed (sin), the calculator displayed 0·5.

90 and then (sin) should give a display of 1.

60 and (cos) should give a display of 0·5.

0 and (cos) should give a display of 1.

45 and (tan) should give a display of 1.

It is more difficult to display 0·5 using (tan), 26·57 gives an answer which is close.

If you found different answers, please discuss them with your teacher because you may be right.

C3

Number	Number squared
20	400
30	900
35	1225
32	**1024**
31	961

C4

Number	Number squared	Number squared $\times$ 4
2	4	16
3	9	36
3·5	12·25	49
2·5	**6·25**	**25**

C5 Your own puzzles.

Challenge

$27 \div 82 = 0.329\,268\,2 \ldots$

Inequations

A1 You've already been given the answer in the booklet.

A2 The box weighs less than 7 kg ($w<7$), but more than 3 kg ($w>3$).

A3

$2w < 9$

A4 Your own working.

A5 (a)

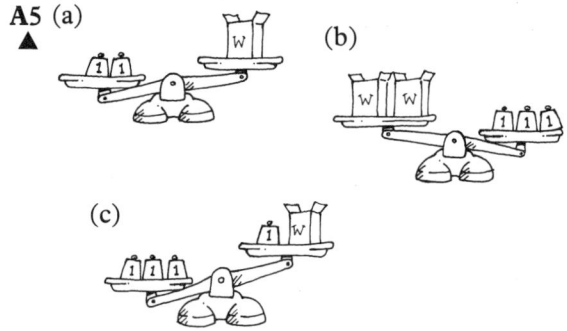

(a) $2 > w$ $2 < w$
(b) $2w < 3$ $3 > 2w$
(c) $3 > 1 + w$ $1 + w < 3$

(a) and (c) each tell us that the weight is less than 2 kg, and (b) tells us that it is less than $1\frac{1}{2}$ kg. It is a whole number so it must be 1 kg.

A6

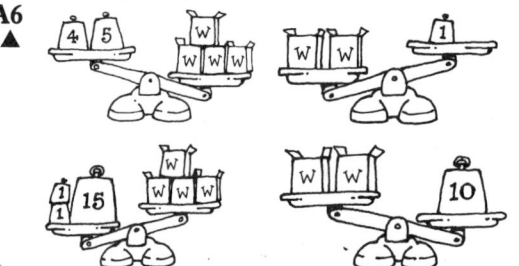

$9 < 4w$ (it helps to write this down as $4w > 9$)

$2w > 1$

$17 > 4w$ (it helps to write this as $4w < 17$)

$2w < 10$

Weight of box w	$2w$	$4w$	$4w>9$	$2w>1$	$4w<17$	$2w<10$
1·5	3	6	No	Yes	Yes	Yes
2	4	8	No	Yes	Yes	Yes
2·5	5	10	Yes	Yes	Yes	Yes
3	6	12	Yes	Yes	Yes	Yes
3·5	7	14	Yes	Yes	Yes	Yes
4	8	16	Yes	Yes	Yes	Yes
4·5	9	18	Yes	Yes	No	Yes

The weights which could be in all four balances are 2·5 kg, 3 kg, 3·5 kg and 4 kg.

A7 Your own inequations.

B1 ▲ (a) $w + 8 < 10$
Take 8 from each side.
$w < 2$
(b) $2w + 4 < w + 6$
Take w from each side.
$w + 4 < 6$
Take 4 from each side.
$w < 2$
(c) $3x + 1 > 2x + 5$
Take $2x$ from each side.
$x + 1 > 5$
Take 1 from each side.
$x > 4$
(d) $5k + 1 > 3k + 11$
Take $3k$ from each side.
$2k + 1 > 11$
Take 1 from each side.
$2k > 10$
$2k$ is greater than 10, so . . .
$k > 5$

Challenge

- Is $k^2 > 0$ always true? What if $k = 0$?
- Is ab always greater than $a+b$?
  Try $a=0\cdot1$ and $b=2$.
- Is $c + d > 0$ always true?
  What if $c=1$ and $d=^-2$?
- Is $r \div s < 1$ always true for any value of r and s? Try $r=10$ and $s=2$.

C1 They could have shown some more results. The report might have been easier to read if they had used paragraphs. What else did you both think?

C2 What do you think?

C3 This table should make the answers easier to understand.

Number of squares (n)	3	4	5	6	7
Smallest perimeter (p)	8	8	10	10	12
Is $p > n + 3$?	yes	yes	yes	yes	yes

Can you see a pattern in this table?
What do you think p would be when $n = 8$?
Would you expect p to be greater than $n + 3$ when $n = 8$?
Check your answers?

C4 Here are some more results – check them against your own.

Number of squares (n)	9	10	11	12	13
Smallest perimeter (p)	12	14	14	14	16
Is $p > n + 3$?	no $p=n+3$	yes	no $p=n+3$	no	no $p=n+3$

C5 ▲ If we label the sides of a triangle like this,

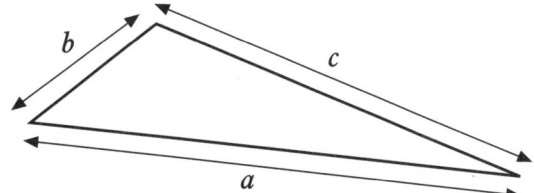

where a is the longest side and b and c are the other sides, $a < b + c$.
In fact, for any triangle, any side is always smaller than the sum of the lengths of the other two sides.

C6 $c < 2r$, where c stands for the length of a chord cutting a circle of radius r.
In fact this is not quite true as you will have seen later on.

C7 (a) $\sqrt{r^2 + s^2} < r + s$ for positive numbers. What about negative numbers?
(b) What if r and s are equal to zero?

C8 $(a + b)^2 > 4ab$, but if a and b are equal then $(a + b)^2 = 4ab$. Try some values of a and b.

C9 The symbol $\geq$ means equal to or greater than.

C10 C6 should be $c \leq 2r$ (A chord could be a diameter.)
C8 $(a + b)^2 \geq 4ab$ ($(a + b)^2 = 4ab$, when a, b are equal)

C11 ▲ $x < 7$　　x is less than 7.
$x \geq 5$　　x is equal to or greater than 5.
The two whole numbers which fit these are 6 and 5.

D1 (a) $2 \leq x$　　x can be any number greater than or equal to 2.
(b) $0 < x < 1$　　x is greater than 0 but less than 1 (it cannot be 0 or 1).
(c) $1 \leq x \leq 1{\cdot}5$　　x can have any value between 1 and 1·5. It can equal 1 or 1·5.
(d) $0 \leq x \leq 0{\cdot}1$　　x can have any value between 0 and 0·1. It can equal 0 or 0·1.

D2 ▲ (a) The first has already been done for you.
(b) *This magnetic disk must not be kept at temperatures above 52 °C or below 10 °C.*
$10\,°C \leq$ temperature $\leq 52\,°C$.
(c) *Unemployed now over 3 million!*
Unemployed $> 3\,000\,000$.
(d) *The meat should be defrosted at maximum power for at least 2 minutes.*
Time at maximum power ≥ 2 minutes.
(e) *Inflation between 10% and 12%.*
$10\% \leq$ inflation $\leq 12\%$.
(f) *In a built-up area a car should not exceed 30 m.p.h.* Speed ≤ 30 m.p.h.

D3 The insect is between 6 mm and 7 mm long (it could be 6 mm or 7 mm long).

A	6 mm	$\leq$	length	$\leq$	7 mm	true
B	6 mm	$>$	length	$<$	7 mm	false
C	7 mm	$\leq$	length	$\geq$	6 mm	false
D	7 mm	$\geq$	length	$\geq$	6 mm	true
E	6 mm	$>$	length	$<$	7 mm	false
F	7 mm	$>$	length	$\leq$	6 mm	false

So B, C, E and F do not fit.

E1

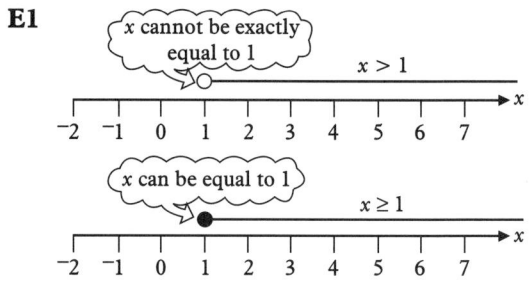

E2

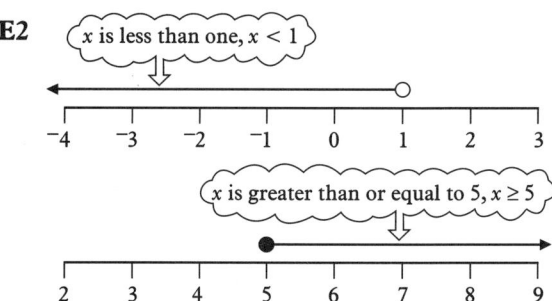

E3 (a) $^{-}1 \leq x \leq 5$　(b) $0 < x < 1$
(c) $0 < x \leq 6$

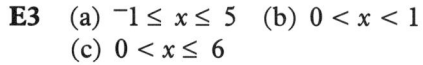

E4 ▲

(a)
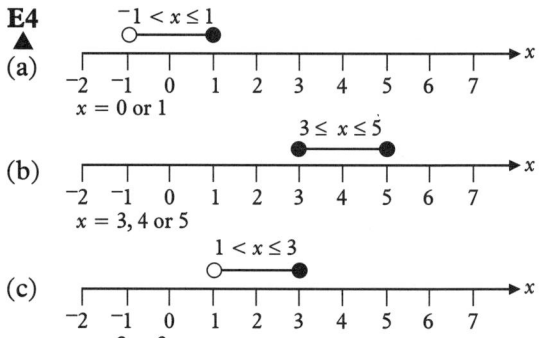
$^{-}1 < x \leq 1$
$x = 0$ or 1

(b)
$3 \leq x \leq 5$
$x = 3, 4$ or 5

(c)
$1 < x \leq 3$
$x = 2$ or 3

(a) $4x < 20$

divide both sides by 5, so $x < 5$.

Dave should have divided by 4, so x < 5.

(b) $x \leq 1$

so x can be 1, 2, 3, 4 and so on.

This is wrong because x is less than or equal to one.

(c) $x < 1$ and $x > {}^-1$, the only whole-number value of x which fits both these inequations is $x = 0$.

These two conditions mean that x is between 1 and ⁻1, but it cannot be either. Dave was right, the only *whole-number* value is x = 0, but there are other possible values such as $x = \frac{1}{2}$.

(d) If $x = {}^-40$ and $y = 0·001$
then $x > y$. **No! look at a number line**

(e)

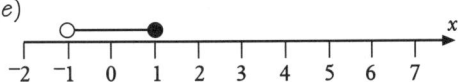

Values of x which fit this are ${}^-1 \leq x \leq 1$
No! ${}^-1 < x \leq 1$

Equations

A1 The length of the spring l is 22,
($3 \times 2 + 16 = 22$).
(You need to find l when $w=2$ in the equation $3w+16=l$.)

A2 For the formula $3w + 16 = l$, when $w=6$
then $l=34$.

A3 For the formula $3w + 16 = l$, let **?** stand for the number of kg (w).
$3(?) + 16 = 49$
$\quad 3(?) = 33$
So $? = 11$
So the weight hanging is 11 kg.

Remember it makes reading a lot easier if each time you have a fresh step of working on a new line. Your working should read like a story.

A4 For the formula $3w + 16 = l$, when $l = 55$:
let **?** stand for the number of kg.
$3(?) + 16 = 55$
$\quad 3(?) = 39$
$\quad$ So $? = 13$
This shows that when $l = 55$, w must be 13 kg.

A5 $3w + 16 = l$, so when $l = 67$
$3w + 16 = 67$
$\quad 3w = 51$
$\quad$ So $w = 17$

A6 Read through your working again if you got any of these wrong.
(a) $w = 18$ (b) $w = 14$.

A7 (a) $s = 32$ (b) $s = 494$.

A8 (a) $30n + 70 = c$
$\quad$ When $n = 5$ miles
$30 \times 5 + 70 = c$
$\quad 150 + 70 = c$
$\qquad$ So $c = 220$p
(b) $30n + 70 = c$
$\quad$ When $c = 430$p
$30n + 70 = 430$
$\qquad 30n = 360$
$\qquad$ So $n = 12$ miles

B1 This is one way to solve the equation.
$\quad 4r - 3 = 25$
Add 3 to each side.
$4r - 3 + 3 = 25 + 3$
$\qquad 4r = 28$
$\qquad$ So $r = 7$
If this is the correct answer (or **solution**) to the equation then when r has a value of 7, $4r - 3$ should equal 25. Does it?

B2 (a) $n = 23$ (b) $n = 14$ (c) $b = 12$
(d) $s = 6$ (e) $t = 11$ (f) $r = 7$
(g) $x = 9$ (h) $y = 9$ (i) $p = 17$
Remember, you should have checked these
for yourselves!

B3 If $n = 8$ is the correct **solution** to the
equation $5n - 16 = 3n$, $5 \times 8 - 16$ should
have the same value as 3×8.
In other words $40 - 16$ should equal 3×8.
Does it?

B4 The correct solutions are
(a) $n = 3$ (b) $n = 5$ (c) $n = 6$
(d) $n = 4$

B5 The correct solutions are
(a) $n = 16$ (b) $n = 3$ (c) $n = 17$
(d) $n = 6$ (e) $n = 7$ (f) $n = 6$

B6 This is one way to solve the equation.
$$5n - 48 = 2n$$
Add 48 to each side.
$$5n - 48 + 48 = 2n + 48$$
$$\text{So } 5n = 2n + 48$$
Subtract $2n$ from each side.
$$\text{So } 3n = 48$$
$$\text{So } n = 16$$

B7 (a) $7n - 32 = 3n$. The solution to this
equation is $n = 8$.
(b) $4n + 33 = 7n$. The solution is $n = 11$.

B8
$$2n + 7 = 5n - 8$$
$$2n + 7 - 7 = 5n - 8 - 7$$
$$2n = 5n - 15$$
$$2n + 15 = 5n - 15 + 15$$
$$2n + 15 = 5n$$
$$15 = 3n$$
$$n = 5$$

B9
$$2n + 7 = 5n - 8$$
$$2n + 7 + 8 = 5n - 8 + 8$$
$$2n + 15 = 5n$$
$$15 = 3n$$
$$n = 5$$

B10 (a) First way
$$5n - 3 + 3 = 4n + 7 + 3$$
$$5n = 4n + 10$$
$$n = 10$$

Second way
$$5n - 3 - 7 = 4n + 7 - 7$$
$$5n - 10 = 4n$$
$$5n - 10 + 10 = 4n + 10$$
$$5n = 4n + 10$$
$$n = 10$$

(b) First way
$$2n + 11 + 5 = 6n - 5 + 5$$
$$2n + 16 = 6n$$
$$16 = 4n$$
$$n = 4$$

Second way
$$2n + 11 - 11 = 6n - 5 - 11$$
$$2n = 6n - 16$$
$$2n + 16 = 6n - 16 + 16$$
$$2n + 16 = 6n$$
$$16 = 4n$$
$$n = 4$$

(c) One way
$$4n - 9 + 9 = 3n + 2 + 9$$
$$4n = 3n + 11$$
$$n = 11$$

Another way
$$4n - 9 - 2 = 3n + 2 - 2$$
$$4n - 11 = 3n$$
$$4n - 11 + 11 = 3n + 11$$
$$4n = 3n + 11$$
$$n = 11$$

C1 These are the last two lines.
$$15n = 5n$$
$$n = 3$$

C2 (a)
$$42 - 2n = 6 + 4n$$
$$42 - 2n + 2n = 6 + 4n + 2n$$
$$42 = 6 + 6n$$
$$36 = 6n$$
$$n = 6$$
(b) $n = 7$ (c) $n = 12$ (d) $n = 9$
(e) $n = 8$

C3 (a) $15 - 3n + 3n = 47 - 11n + 3n$
$$15 = 47 - 8n$$
$$15 + 8n = 47 - 8n + 8n$$
$$15 + 8n = 47$$
$$8n = 32$$
$$n = 4$$

(b) $15 - 3n + 11n = 47 - 11n + 11n$
$$15 + 8n = 47$$
$$8n = 32$$
$$n = 4$$
(c) Method (b) is probably easier.

C4 (a) $a = 5$ (b) $b = 6$ (c) $n = 5$
(d) $n = 3$

Number searches

A better name for this section is 'trial and improvement'. You should get better after each trial not worse!

A1 (a) 8 cm by 4 cm   (b) 12 cm by 24 cm
(c) 6 cm by 12 cm   (d) 20 cm by 40 cm
(e) 16 cm by 32 cm   (f) 15 cm by 30 cm

A2 (a) 9 cm by 8 cm   (b) 13 cm by 12 cm
(c) 21 cm by 20 cm   (d) 16 cm by 15 cm
(e) 25 cm by 24 cm   (f) 32 cm by 31 cm

A3 70 cm by 35 cm and 50 cm by 49 cm

A4 (a) 10 cm by 10 cm   (b) 13 cm by 13 cm
(c) 31 cm by 31 cm   (d) 44 cm by 44 cm
(e) 70 cm by 70 cm   (f) 74 cm by 74 cm
(g) 77 cm by 77 cm   (h) 86 cm by 86 cm

A5 256 makes a 16 cm square,
900 makes a 30 cm square and 1600 makes a 40 cm square. With questions like these it sometimes helps to use a table.

B1 (a) 34 cm
(b) 1122 sq. cm ($33 \times 34 = 1122$)

B2 You may have used a table like this one to help you find the rectangle with an area of 650 sq. cm.

Height (cm)	Width (cm)	Area (sq. cm)
20	21	420 too small
30	31	930 too large
23	24	552 too small
25	26	650 spot on!

So the correct answer is a rectangle which is 26 cm by 25 cm.

B3 A table may come in handy here. It helps you see how to narrow down your search.

h (cm)	$h + 1$ (cm)	$h \times (h + 1)$
50	51	2550
90	91	8190
81	82	6642
87	88	7656
83	84	6972
84	85	7140

The correct answer for the value of h which gives a value of 7140 for $h \times (h + 1)$ is $h = 84$. In fact this person could have found the answer a little quicker because $h \times (h+1)$ ended in a zero. This means that h or $h+1$ must end in a zero or a five. Why?

B4 $h = 53$ **B5** $h = 68$ **B6** $h = 30$

B7 See the next section.

C1 Your diagram should look like this.

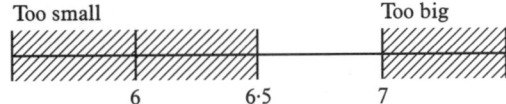

C2 This is what your diagram should look like.

C3

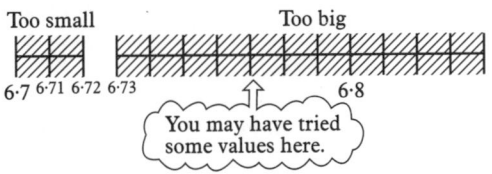

Look carefully at this diagram to find the answers to (a), (b) and (c).

(d) *h lies between 6·72 and 6·73.*

C4 (a) *h lies between 6·728 and 6·729.*
▲ (b) *In fact you cannot find h exactly.*

C5 *The height is between 2·70 m and 2·71 m.*

C6 The height is between 2·31 m and 2·32 m.

19

C7 (a) x is between 8·19 cm and 8·20 cm.
    (b) x is between 12·28 cm and 12·29 cm.

D1

Length of side in cm	0	1	2	3	4	5	6	7	8
Area in sq. cm	0	1	4	9	16	25	36	49	64

(b) and (e)

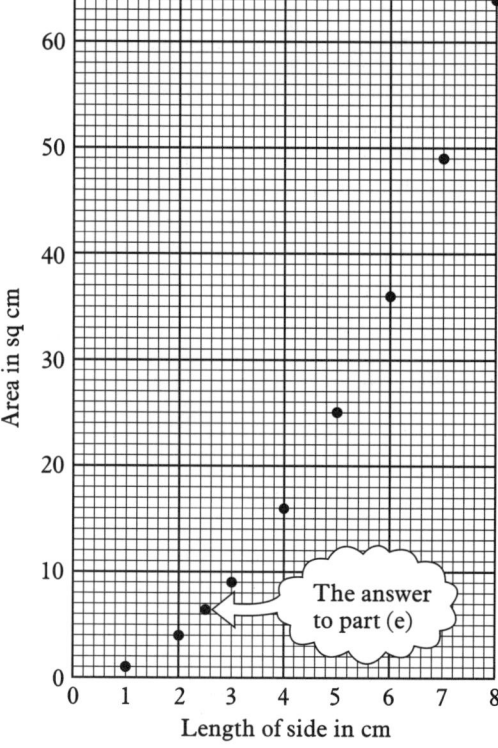

The answer to part (e)

(c) The area of a square with sides 2·5 cm is 6·25 sq. cm (or cm²).

(d) No, the area halfway between the areas of squares of side 2 cm and 3 cm would be 6·5 cm².

> Your answers to **D2**, **D3** and **D4** may be slightly different from these.

D2 (a) 23 cm² (b) 10 cm² (c) 35 cm²

D3 (a) 5·9 cm (b) 4·5 cm (c) 5·5 cm

D4 (a) Between 16 cm² and 64 cm².
    (b) Between about 7 cm² and 29 cm².
    (c) Between about 42 cm² and 56 cm².

D5 (a) Between 7 cm and 8 cm.
    (b) Between 5 cm and 7 cm.
    (c) Between 6 cm and 10 cm.

D6 Between 5 cm and 6 cm.

D7 Between 9 cm and 10 cm.

D8 Look at your own table.
The length of the side is between 3·162 cm and 3·163 cm.

D9 The length of the side is between 4·123 cm and 4·124 cm.

D10 Between 5·477 cm and 5·478 cm.

D11 The sides of a square which has an area of 17 cm² are 4·12 cm correct to 2 decimal places.

D12 If a square has an area of 30 cm², then its sides correct to 2 decimal places are 5·48 cm.

E1 (a) 5 is the square root of 25 ($5 \times 5 = 25$).
    (b) 2 is the square root of 4.
    (c) 8 is the square root of 64.
    (d) 4 is the square root of 16.

E2 The square root of 7 ($\sqrt{7}$) lies between 2·64 and 2·65.

E3 The square root of 20 ($\sqrt{20}$) lies between 4·47 and 4·48.

E4 $\sqrt{50}$ lies between 7·071 and 7·072, so it is 7·07 to 2 decimal places.

Expressions

Expressions with a calculator

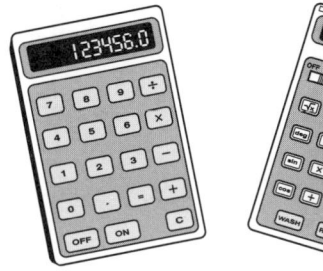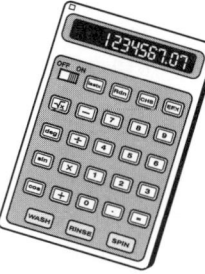

Investigate

Some possible ways to make 9 are:

| 2 | + | 7 | = |

| 3 | × | 3 | = |

| 8 | 1 | √ |

You should be able to find others.

A1–A4 The answers to these will depend on the type and make of your calculator. Check your own answers.

There are six possibilities using two of the numbers. When all three numbers are used, there are three possibilities with a scientific calculator and six with a four-function calculator.

A5 $5 + 10 \times 100 =$ will give the largest number. This will be 1500 or 1005 depending on your calculator.

Challenge

Two examples are:

$3 \times 5 + 3 = 18$ which gives the same answer as $3 + 5 \times 3 = 18$

and $4 \times 1 + 7 = 11$ which gives the same answer as $4 + 1 \times 7 = 11$.

A6 (a) Sandra got the same answer 12 for both calculations.
(b) She used a four-function calculator.

A7 (a) A scientific calculator was used for the calculation $5 + 3 \times 7 = 26$. A four-function calculator would have given the answer 56.
(b) A four-function calculator was used. A scientific calculator would have given the answer 11.
(c) A four-function calculator was used. A scientific calculator would have given the answer 14.
(d) A four-function calculator was used. A scientific calculator would have given the answer 5.

A8 You are likely to find that a four-function calculator works out the answers as it goes along and gives:

this gives 8
$\overbrace{4 \times 2} - 3 = 5$

this gives 2
$\overbrace{4 - 2} \times 3 = 6$

A scientific calculator works out the multiplications first and gives:

this gives 8
$\overbrace{4 \times 2} - 3 = 5$

this gives 6
$4 - \overbrace{2 \times 3} = -2.$

A9 (a) $5 + 4 + 3 = 12$ on any calculator
(b) $5 \times 4 - 3 = 17$ on any calculator
(c) $5 + 4 \times 3 = 27$ on a four-function calculator
(d) $5 \times 4 \times 3 = 60$ on any calculator

There are many different ways to obtain some of these numbers. Possible answers are:

$0 = 3 + 3 - 3 - 3$
$1 = 3 \times 3 \div 3 \div 3$
$2 = 3 \div 3 + 3 \div 3$ (scientific calculator only)
$3 = 3 + 3 + 3 \div 3$ (four-function only)
$4 = 3 \times 3 + 3 \div 3$ (four-function only)
$5 = 3 + 3 - 3 \div 3$ (scientific calculator only)
$6 = 3 - 3 + 3 + 3$
$7 = 3 \div 3 + 3 + 3$
$8 = 33 \div 3 - 3$
$9 = 3 \times 3 - 3 + 3$
$10 = 3 \times 3 + 3 \div 3$ (scientific calculator only)

Try this

You should find that the four-function calculator does the calculations in the order in which you enter them. The scientific calculator works out the divisions and multiplications before it does the additions and subtractions.

B1 Both are right in a way. See page 8 of the booklet.

B2 Alan is starting from the left and so does the additions before the multiplications. Liz does the multiplications first no matter where they occur.

B3 Remember, for expressions like these the 'scientific calculator' method is used. In other words we always do the multiplications or divisions before any additions or subtractions.
(a) 23   (b) 2   (c) 20   (d) 18
(e) 20   (f) 26

B4 Work out $6 \cdot 9 \times 11 \cdot 9 = 82 \cdot 11$
▲ $82 \cdot 11 + 1 \cdot 74 = 83 \cdot 85$

B5 Here are just two, you may have found some more.
$$6 - 2 \qquad (6 - 4) \times 2$$

C1 (a) $(7 + 9) \times 5 = 80$ (b) $(9 - 7) \times 4 = 8$
▲   (c) $(5 + 7 + 12) \div 6 = 4$

C2 (a) $(2 + 3) \times 7$ (b) $(4 \cdot 2 \times 5) + 3 \cdot 5$
(c) $(4 \times 29) + 85$ (d) $(90 + 6 \times 45) \times 4$
(e) $(3 \times 56) + (2 \times 38) + 85$

Talking Point

It is wrong to write $24 \times 5 = 120 + 8$ because the left-hand side makes 120 and the right-hand side makes 128.
The two sides are not equal.

D1 (a) $(4 \times 10) + (2 \times 100) + (3 \times 1000)$
$= 40 + 200 + 3000$
$= 3240$
(b) see (a) above.
(c) $2010 = (2 \times 1000) + 10$
$= \cap \pounds \pounds$

D2 Here are two more values you can get using brackets:
$8 \times (4 \div 4) \times (2 - 1) = 8$
$8 \times (4 \div 4) \times 2 - 1 = 15$

D3 You could ask someone else to try finding different values of your expression using brackets.

D4 The pattern is   7   9   11   13   15   **17**

D5 All the numbers are odd – you may have noticed other things.

D6 $7 = (1 \times 2) + 5$
  $9 = (2 \times 2) + 5$
  $11 = (3 \times 2) + 5$
  $13 = (4 \times 2) + 5$
  and so on.
  (a) The 10th number is $(10 \times 2) + 5 = 25$.
  (b) The 30th number is $(30 \times 2) + 5 = 65$.

D7 $5 = (1 \times 4) + 1$
  $9 = (2 \times 4) + 1$
  $13 = (3 \times 4) + 1$
  The 11th number in the pattern is
  $(11 \times 4) + 1 = 45$.

D8 $4 = (1 \times 3) + 1$
  $7 = (2 \times 3) + 1$
  $10 = (3 \times 3) + 1$
  $13 = (4 \times 3) + 1$
  $16 = (5 \times 3) + 1$
  Look carefully at the pattern. The 12th
  number is $(12 \times 3) + 1 = 36 + 1$
  $\qquad\qquad\qquad\qquad\quad = 37$

D9 $1 = 1 \times (1 + 1) \div 2$
  $3 = 2 \times (2 + 1) \div 2$
  $6 = 3 \times (3 + 1) \div 2$
  $10 = 4 \times (4 + 1) \div 2$
  The 7th number is
  $7 \times (7 + 1) \div 2 = 7 \times (8) \div 2 = 7 \times 4 = 28$.
  This pattern is the triangle numbers.

D10 $5 = (1 \times 6) - 1$
  $11 = (2 \times 6) - 1$
  $17 = (3 \times 6) - 1$
  $23 = (4 \times 6) - 1$
  The next number is $(5 \times 6) - 1 = 29$.
  All these numbers are odd numbers. They
  are also prime numbers. Are the next few
  numbers in the pattern prime numbers as
  well?

Challenges

1 (a) $(2 + 3) \times (10 - 3) = 5 \times 7 = 35$
   (b) $(8 + 2 \times (4 + 6)) \times 2 =$
      $(8 + 2 \times 10) \times 2 = 28 \times 2 = 56$
   (c) $(10 - (2 + 5)) \times 3 =$
      $(10 - 7) \times 3 = 9$
   (d) $(4 \times 20) \div (2 \times 5) = 80 \div 10 = 8$

Expressions with letters

A1 You probably found (c) the easiest because it had all the items grouped.

A2 (a) Mike spent 5×35 pence on baked beans.
  (b) He spent $5 \times 35 = 175$ or £1·75 on baked beans.
  (c) An expression for the total amount (in pence) Mike spent on pet food is $7 \times 31 + 2 \times 28$. (You may also have written $2 \times 28 + 7 \times 31$.)

A3 (a) The total Lorna spent on cat food at the Hypermarket was 7×31 (pence)
  (b) This comes to 217p or £2·17.
  (c) If cat food was reduced by 5p a tin the new expression would be 7×26 (p). This gives a total of 182p or £1·82.
  (d) Did you think about "travel costs", "lack of choice", etc.?

B1

Four short pipes have a total length of 4×3 metres. This is 12 metres.

B2 For four short and five long pipes, the total length is $4 \times 3 + 5 \times 8$ metres.

Investigate

- There are only two ways of making up a total length of piping of 40 metres. 5 lots of 8 m lengths ($5 \times 8 = 40$) or 2 lots of 8 m lengths and 8 lots of 3 m lengths ($2 \times 8 + 8 \times 3 = 16 + 24 = 40$)

- Here are all the lengths up to 200 metres it is impossible to make without cutting the 8 m or 3 m lengths of pipe: 1, 2, 4, 5, 7, 10 and 13. This may surprise you.
  If you think there are any more check your results again.
  This might be a deliberate mistake!

B3 (a) The total length of 6 pipes each of 2 metres length is 2×6 metres.
(b) For p pipes the total length is $2 \times p$, but it's easier to write $2p$. You might have written $p \times 2$ or $p2$, but we usually put the number first.

B4 Toni is right but Salik and Debra have made simple, easy to make, mistakes. Debra has muddled up 'm' meaning metres and 'm' which could stand for a number. Salik has got the number of pipes, n, correct, but forgotten to multiply this by 10 to get the total length.

B5

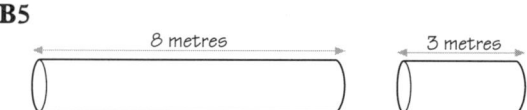

An expression for the total length of 2 short and 5 long pipes is $2 \times 3 + 5 \times 8$ (metres)

B6 For 7 short and 2 long the expression is $7 \times 3 + 2 \times 8$.

B7 For a short and b long the expression is $3a + 8b$.

B8 The total length of x 2-metre pipes, y 3-metre pipes and z 5-metre pipes is $2x + 3y + 5z$.

C1

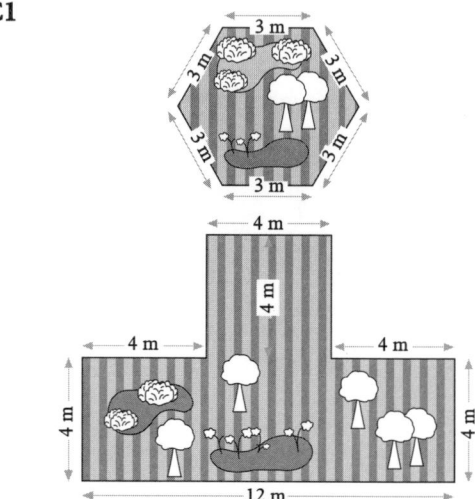

(a) The perimeter is $6 \times 3 = 18$ metres
(b) This garden has a perimeter of $7 \times 4 + 12 = 40$ metres.

C2 (a) This garden needs $6k$ metres of fencing round its perimeter.
(b) The perimeter of this garden is $12g$ metres.

C3 The perimeters in metres are:
(a) $2g + 3h$ (or $3h + 2g$)
(b) $4x + 4y$ (or $4y + 4x$)
(c) $6r$ ($2r + r + 2r + r$)
(d) $4w + 6v$
(e) $5q$ (a pentagon has five sides)
(f) $8y$ (4 lots of $2y$)

C4 If you are not sure of any of these, it can sometimes help if you write down the expression using simple numbers before using letters.
(a) $? = y - x$ (b) $? = w - v$
(c) $? = e - f$ (d) $? = p + q$

C5

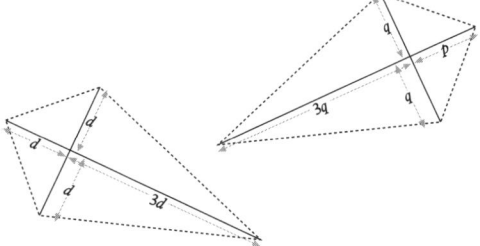

(a) Length of wood needed is $6d$.
(b) Length of wood needed is $5q + p$.

C6

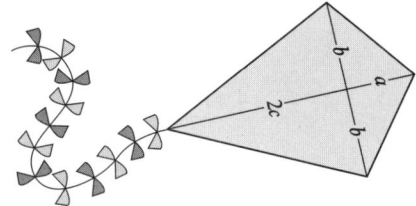

This kite needs $a + 2b + 2c$ metres of wood. You may have written $2b + 2c + a$. This is correct, but when there are several different letters we usually put them in alphabetical order.

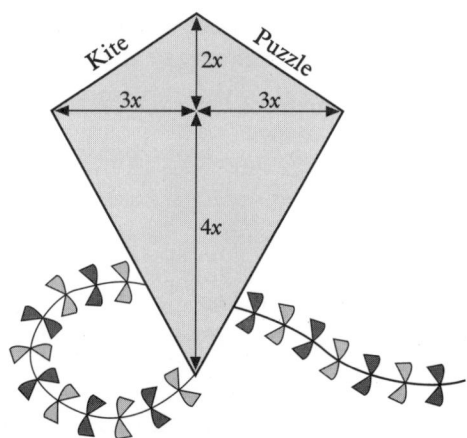

This is one way, you may have found others. The important thing is that the total wood used must be $12x$ metres.

C7

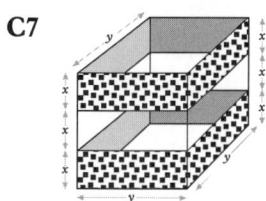

This kite will need $12x + 16y$ centimetres of wire.

$(3x + 3x + 3x + 3x + 4y + 4y + 4y + 4y)$

Challenges

- Each of the eight panels has an area of $x \times y$ square centimetres.
  We can write this as xy square centimetres.
  So the total area is $8xy$ square centimetres.

- To do this you need to know how to find the area of a triangle.
  The answer is $ab + 2bc$ square centimetres.
  You will need to ask your teacher if you were stuck.

D1 The width of the paper is $2f$ or $3a$.
The length of the paper is b or $c + d + e + 2$.
Look back to page 14 if you are still unsure.

D2 The expression for the distance x is e, and for the distance marked y it is c.

D3 Did it work!

Investigate

What did you find?

E1 If you're uncertain of any of these ask your teacher.

Expression	Correct answer	
$2c + c$	$3c$	
$2a + b + a$	$3a + b$	
$a + 2b$	$a + 2b$	This expression can't be simplified.
$a + a + a$	$3a$	We put the number first.
$2x + y$	$2x + y$	It can't be simplified any more.
$a + a + 2a$	$4a$	
$2a + b + a + 2b$	$3a + 3b$	
$1 + a + 1$	$a + 2$ or $2 + a$	
$4a - a$	$3a$	

Expressions

A1 At the start there were 740 cans, 230 more were delivered and 410 sold. So at the end of the week there were $740 + 230 - 410$ cans left. The value of this expression is 560.

A2 For this question you need to work out this calculation.
$3800 + 1200 - 1700 = 3300$.

A3 (a) The order the manager does the calculation does not matter.
(b) This expression gives the number of cans left at the end of the week:
$b - s + d$.
In this expression b stands for the number at the beginning of a week, d for the number of cans delivered and s for the number sold.

A4 These two expressions also give the number of cans left at the end of a week.
(a) $d - s + b$ and (d) $d + b - s$
If you are not convinced try some expressions with numbers in before using letters.

A5 $b - s + d$ (as well as $d - s + b$ and $d + b - s$ which were the answers to question **A4**).

A6 (a) This expression gives the number of pence Rajesh has left $g - s - c - f$,
where g stands for the amount he starts with
s shows how much he spends on sweets.
c how much he spends on comics
and f how much he spends on fares.
All these are in pence.
(b) Any of these expressions are equivalent to the one in (a):
$$g - s - f - c$$
$$g - c - s - f$$
$$g - c - f - s$$
$$g - f - s - c$$
$$g - f - c - s$$

> ### EQUIVALENT EXPRESSIONS
> The expressions $b + d - s$ and $d + b - s$ always give the same result, no matter what numbers b, d, s stand for.
>
> They are called **equivalent expressions**.

B1 (a) Yes, the answer is the same as before.
(b) Your answer should be the same.

B2 (a) and (c) give the same answer as before.

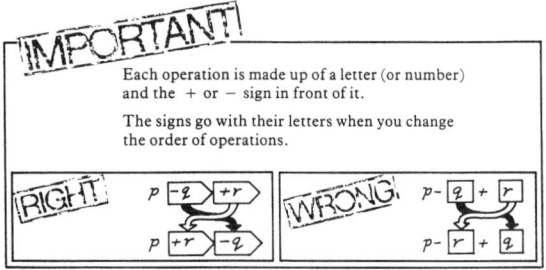

IMPORTANT!
Each operation is made up of a letter (or number) and the + or − sign in front of it.
The signs go with their letters when you change the order of operations.
RIGHT
WRONG

B3 (b) and (d) are equivalent to $w - x - y + z$. In other words, $w + z - y - x$ and $w - y + z - x$ are equivalent.

B4 All these expressions are equivalent. (They do not include the ones you have seen already.)
$w - x + z - y \quad w - y - x + z \quad w + z - x - y$

C1 (a) Both expressions have a value of 40.
(b) Both expressions have a value of 68 when $n = 9$.

C2 (a) The expression $4n - 2 - 5 + 2n$ is simplified to $6n - 7$.
(b) Both expressions have a value of 53 when $n = 10$.

C3 (a) $3n + 5$ (b) $8n + 5$ (c) $12n - 10$
(d) $7n + 2$ (e) $8n - 6$ (f) $8n + 5$

C4 (a) $9n + 7$ (b) $2n + 3$ (c) $2n + 10$
(d) $14 - 5n$ (e) $4n + 2$ (f) $n + 4$

C5 (a) $5n - 7$ (b) $2n + 11$ (c) $2n + 2$
(d) $1 - 4n$ (e) $5 - 2n$ (f) $3n + 5$

C6 $6n - 2$

C7 (a) $2n - 4$ (b) $5 - 2n$

D1 (a) $n = 4$ (b) Your own check (the left-hand side does work out to be 10).

D2 (a) $n = 2$ (b) $n = 3$ (c) $n = 3$
(d) $n = 3$ (e) $n = 2$ (f) $n = 4$

D3 (a) $n = 2$ (b) $n = 1$ (c) $n = 4$
(d) $n = 4$

Brackets 1

Here is a picture of $(3 \times 5) + 2$. There are 3 'bags' of 5 dots and 2 extra dots. (3×5) + 2	Here is a picture of $3 \times (5 + 2)$. There are 3 bags, each containing $5 + 2$. $3 \times (5 + 2)$

This is another way of grouping the dots in the second picture. It shows that

$3 \times (5 + 2) = (3 \times 5) + (3 \times 2)$.

$(3 \times 5) + (\ \times 2)$

A1 (a) $5 \times (4 + 7) = (5 \times 4) + (5 \times 7)$
(b) $6 \times (9 + 2) = (6 \times 9) + (6 \times 2)$
(c) $8 \times (5 + 3) = (8 \times 5) + (8 \times 3)$

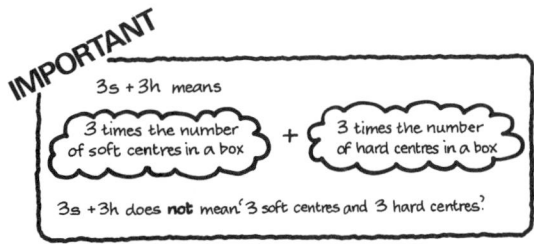

IMPORTANT

$3s + 3h$ means

3 times the number of soft centres in a box + 3 times the number of hard centres in a box

$3s + 3h$ does **not** mean '3 soft centres and 3 hard centres.'

A2 (a) $s = 15, h = 8$. This means that a box of 'Silver Leaf' has 15 soft centres and 8 hard centres.
(b) The value of $3(s + h)$ for 'Silver Leaf' is $3(15 + 8)$ which is 69.

A3 The expression $4(s + h)$ gives the total number of chocolates in four boxes of 'White Magic'. For 'White Magic' $s = 12$ and $h = 10$. You can work out the total number of chocolates in the four boxes. There are two ways of doing this:
$$4(s + h) = 4(12 + 10) = 88$$
or $\quad 4s + 4h = 4 \times 12 + 4 \times 10$
$$= 48 + 40$$
$$= 88.$$

A4 (a) 120 $\quad [6 \times (7 + 13)$ or $6 \times 7 + 6 \times 13]$
(b) 180 $\quad [10 \times (9 + 9)$ or $10 \times 9 + 10 \times 9]$
(c) 180 $\quad$ [see above]

A5 (a) $5(s + h) = 5s + 5h$
(b) $7(s + h) = 7s + 7h$
(c) $8(p + q) = 8p + 8q$
(d) $2(p + q + r) = 2p + 2q + 2r$
(e) $5(w + x + y + z) = 5w + 5x + 5y + 5z$

A6 (a) $2(p + 2) = 2p + 4$
(b) $7(4 + y) = 28 + 7y$
(c) $3(2 + p + q) = 6 + 3p + 3q$
(d) $5(x + 4) = 5x + 20$
(e) $4(n + 5) = 4n + 20$
(f) $2(x + 1 \cdot 5) = 2x + 3$

A7 (a) $4 \times (5 - 2) = 12$
(b) $4 \times (5 - 2) = (4 \times 5) - (4 \times 2)$
a minus sign

A8 (a) $3(a - b) = 3a - 3b$

(b) Whichever way you use the answer is 6.
$[3(7 - 5) = 3 \times 2 = 6$ or
$3 \times 7 - 3 \times 5 = 21 - 15 = 6]$

A9 (a) $5x - 5y$ (b) $4a - 4b$ (c) $7p - 7q$
     (d) $5x - 10$ (e) $24 - 8x$ (f) $20 - 4m$
     (g) $3p + 3q - 3r$ (h) $2s - 2t + 2u$
     (i) $28 - 4p + 4q$ (j) $2a - 6 + 2b$
     (k) $5a - 15 - 5b$ (l) $2l + 12 - 2m$

B1

(a) Check your table against this one.

Number of litres poured from bucket into barrel	Number of litres left in bucket	Number of litres in barrel
0	13	17
1	12	18
2	11	19
3	10	20
4	9	21
5	8	22
6	7	23
7	6	24
8	5	25

(b) A row where the amount in the barrel is exactly 3 times the amount left in the bucket does not appear in the table.

B2
$$17 + p = 3(13 - p)$$
$$17 + p = 39 - 3p$$
$$17 + p + 3p = 39 - 3p + 3p$$
$$17 + 4p = 39$$
$$4p = 22$$
$$p = 5{\cdot}5$$

B3 There are two strips of metal. Their lengths are 13 cm and 23 cm. A piece is cut off the short strip and welded onto the long strip. Now the long strip is 5 times as long as the short strip.
Let c cm be the length cut off.
(a) The length of the short piece is now $13 - c$.
(b) The length of the long piece is now $23 + c$.
(c) This equation shows that the long piece is now five times as long as the short one: $23 + c = 5(13 - c)$.
Make sure you have the '5' on the correct side.

(d) The solution to the equation is $c = 7$.
(e) Your own check

B4 The equation is $19 + g = 3(15 - g)$, leading to $g = 6{\cdot}5$.
So John gave Sally £6·50.

B5 Let w be the weight of wheat poured.
So $4(35 - w) = 25 + w$
The solution to this equation is 23.
So 23 kg of wheat was poured.

B6 ▲ Jack has £20. Jill has £9. They each spend the same amount of money.
After this Jack has 3 times as much as Jill.
Let s be the number of pounds each person spent.
(a) Jack has $20 - s$ left.
(b) Jill has $9 - s$ left.
(c) This equation tells you that Jack has three times as much as Jill:
$$20 - s = 3(9 - s)$$
(d) The solution to the equation is $s = 3{\cdot}5$
So they each spend £3·50.
(e) Your own check

B7 Let x be the amount of water poured into each container.
(You can use any letter you wish.)
The bucket now contains $3 + x$ litres of water and the barrel $36 + x$ litres.
The barrel contains 7 times as much water as the bucket.
The equation for this is $36 + x = 7(3 + x)$
This leads to $x = 2{\cdot}5$.
So 2·5 litres were poured into each container.

B8 Peter is 34 years old now, and his son Jason is 8.
(a) n years from now Peter will be $34 + n$.
(b) Jason will be $8 + n$, in n years time.
(c) n years from now Peter will be 3 times as old as Jason:
$$34 + n = 3(8 + n).$$
(d) The solution to the equation is $n = 5$ (check this solution yourselves).

B9 (a) $29 + n = 4(2 + n)$, leading to $n = 7$.
So Mary will be 4 times as old as Jane in 7 years time.
(b) In 18 years.

B10 (a) $x = 21$ (b) $x = 9$ (c) $x = 6$
(d) $x = 3$

B11 (a) $x = 7$ (b) $x = 2$ (c) $x = 3$
(d) $x = 0{\cdot}5$

B12

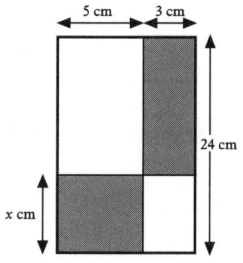

If the two shaded parts of the rectangle have the same area, then
$$5x = 3(24 - x).$$
So $x = 9$ cm. Check that this is true.

Brackets 2

Add 35p and 28p. Subtract the total from 80p, like this.
$80 - (35 + 28)$

Start with 80p. First subtract 35p, then 28p, like this
$80 - 35 - 28$

A1 (a) $20 - (4 + 7) = 20 - 4 - 7$
(b) $94 - (16 + 47) = 94 - 16 - 47$
(c) $p - (q + r) = p - q - r$
(d) $5 - (s + t) = 5 - s - t$
(e) $w - (3 + x) = w - 3 - x$
(f) $y - (z + 2) = y - z - 2$

A2 (a) $10 - (7 - 3) = 6$
(b) $10 - (7 - 3) = 10 - 7 + 3$

A3 (a) $12 - (4 - 1) = 12 - 4 + 1$
(b) $6 - (3 - 2) = 6 - 3 + 2$
(c) $14 - (9 - 3) = 14 - 9 + 3$
(d) $12 - (8 - 5) = 12 - 8 + 5$

A4 $h - (s - t) = h - s + t$

A5 Your own explanation – does it still make sense to you now?

A6 (a) $a - (b - c) = a - b + c$
(b) $p - (q - 2) = p - q + 2$
(c) $r - (5 - s) = r - 5 + s$
(d) $10 - (f - g) = 10 - f + g$

A7 (a) 10
(b) $20 - (9 + 4 - 3) = 20 - 9 - 4 + 3$

A8 (a) $10 - (3 + 2 + 1) = 10 - 3 - 2 - 1$
(b) $12 - (7 - 3 - 2) = 12 - 7 + 3 + 2$
(c) $15 - (8 + 4 - 1) = 15 - 8 - 4 + 1$
(d) $20 - (6 - 2 + 5) = 20 - 6 + 2 - 5$
(e) $8 - (2 + 4 - 3) = 8 - 2 - 4 + 3$

┌─ **First rule for removing brackets** ─

When there is a **subtract** sign in front of the brackets, every **add** sign **inside** the brackets changes to **subtract**, every **subtract** sign **inside** the brackets changes to **add**.

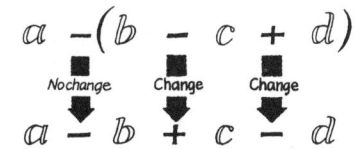

$a - (b - c + d)$

No change — Change — Change

$= a - b + c - d$

A9 (a) $a - b - c + d$ (b) $w - x - y - z$
(c) $p - q + r + 3$ (d) $a - 2 + b - c$
(e) $4 - w + x - y$ (f) $m - n + p - 1$
(g) $f - g + h - i - j$ (h) $r - 2 - s - t + u$

A10 $a - b + c - (d - e + f) = a - b + c - d + e - f$

A11 (a) $p + q - r + s$ (b) $s - t - u - v$
(c) $a + b - c + d - e$ (d) $a - b - c - d + e + f$

┌─ **Second rule for removing brackets** ─

When there is an **add** sign in front of the brackets, the **add** and **subtract** signs **inside** the brackets **do not change**.

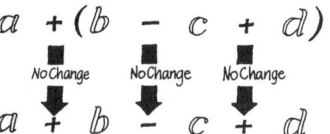

$a + (b - c + d)$

No Change — No Change — No Change

$a + b - c + d$

B1 (a) $7 + (5 + 4) = 16$
(b) $7 + (5 + 4) = 7 + 5 + 4$
(c) $12 + (8 + 3) = 12 + 8 + 3$
(d) $a + (b + c) = a + b + c$

B2 (a) $10 + (7 - 2) = 15$
(b) $10 + (7 - 2) = 10 + 7 - 2$
(c) $16 + (9 - 4) = 16 + 9 - 4$
(d) $a + (b - c) = a + b - c$
(e) $20 + (8 - 3 + 2) = 20 + 8 - 3 + 2$
(f) $10 + (6 + 3 - 4) = 10 + 6 + 3 - 4$
(g) $a + (b - c + d) = a + b - c + d$
(h) $a + (b + c - d) = a + b + c - d$

B3 (a) $p + q - r - s$ (b) $w + x + y - z$
(c) $a + b + c - d - e$ (d) $f - g + h - i + j$
(e) $k + l + m + n - p + q$

B4 (a) $p - q - r$ (b) $f - g + h$ (c) $s + t - u$
▲ (d) $l + m + n$ (e) $x - y - 3$ (f) $h + 7 = k$
(g) $9 - a - b$ (h) $7 + a - b$
(i) $p - q + r - s + t$ (j) $a + b - c - d + e$

B5 (a) $p + q + r - s - t + u$
(b) $h - i - j - k + l - m - n$
(c) $a - b - c + d + e + f - g + h$
(d) $s - t + u + v + w - x + y$

B6 (a) $a - (b + c)$ (b) $p + (q - r)$
(c) $s + (t + u)$ (d) $f - (g - h)$
(e) $a - (b + c - d)$ (f) $x - (y - z + 5)$

C1 (a) $13 - x$ (b) $4 + x$ (c) $5 + x$
(d) $7 + a$ (e) $14 + a$ (f) $13 - a$
(g) $5 + p$ (h) $7 - p$ (i) $5 - p$
(j) $2 + x$ (k) $3 - y$ (l) $9 - x$

C2
$7x - (2 - 3x)$ Rough working
$= 7x - 2 + 3x$ $7x + 3x$
$= 10x - 2$

C3 (a) $4a + (2a - 3) = 6a - 3$
▲ (b) $8p - (3p + 2) = 5p - 2$
(c) $4x + 3$ (d) $12y + 3$
(e) $5b - 4$ (f) $3c + 7$

C4 (a) $6a - 7$ (b) $3x + 6$ (c) $5p - 1$
(d) $2n + 5$ (e) $5y + 6$ (f) $19 - 6q$

C5 (a) $8 - (3 + 2p) - 3p - 4 = 1 - 5p$
(b) $10 - (2a - 3) + (5a - 2) = 11 + 3a$
(c) $4n - 3 - (n - 1) + 5 = 3n + 3$
(d) $x - 4 - (2x - 3) - 6 = {}^-x - 7$

C6 (a) $7a - 3$ (b) Right   (c) Right
(d) $3y - 2$ (e) Right   (f) $10 + a$
(g) Right   (h) $13a - 5$ (i) $11 + 3a$

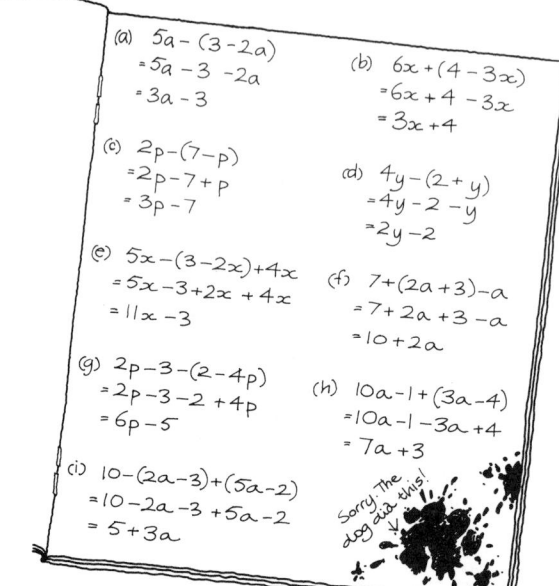

(a) $5a - (3 - 2a)$
$= 5a - 3 - 2a$
$= 3a - 3$

(b) $6x + (4 - 3x)$
$= 6x + 4 - 3x$
$= 3x + 4$

(c) $2p - (7 - p)$
$= 2p - 7 + p$
$= 3p - 7$

(d) $4y - (2 + y)$
$= 4y - 2 - y$
$= 2y - 2$

(e) $5x - (3 - 2x) + 4x$
$= 5x - 3 + 2x + 4x$
$= 11x - 3$

(f) $7 + (2a + 3) - a$
$= 7 + 2a + 3 - a$
$= 10 + 2a$

(g) $2p - 3 - (2 - 4p)$
$= 2p - 3 - 2 + 4p$
$= 6p - 5$

(h) $10a - 1 + (3a - 4)$
$= 10a - 1 - 3a + 4$
$= 7a + 3$

(i) $10 - (2a - 3) + (5a - 2)$
$= 10 - 2a - 3 + 5a - 2$
$= 5 + 3a$

Sorry the dog did this!

D1 (a)
$7n - (3 + 2n) = 12$ Rough working
$7n - 3 - 2n = 12$ $7n - 2n$
$5n - 3 = 12$ -3
$5n = 15$
so $n = 3$

(b) Your own check

D2 Don't forget you should have checked these answers already!
(a) $n = 2$ (b) $n = 4$ (c) $n = 3$ (d) $n = 4$
(e) $n = 2$ (f) $n = 3$ (g) $n = 1$ (h) $n = 11$

E1 (a)

	9	
4	19	23
	28	

(b)

	1	
6	17	23
	18	

(c)

	8	
2	9	11
	17	

(d)

	4	
4	12	16
	16	

(e)

	2	
7	11	18
	13	

(f)

	0	
8	15	23
	15	

E2 The solution to $34 + n = 3n$ is $n = 17$.
▲

E3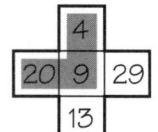

E4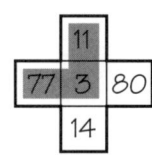

E5 Look back at the original question.

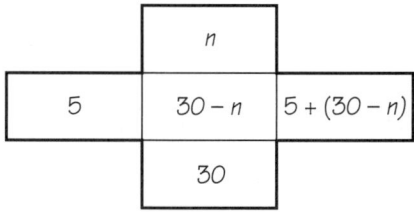

The right-hand number
is 4 times the top number
so
$5 + (30 - n) = 4n$.
The solution to this equation is $n = 7$.
Don't forget to check that $n = 7$
fits the puzzle!

(a) $30 - n$ (b) $5 + (30 - n)$
(c) $5 + (30 - n) = 4n$ (d) $n = 7$

E6 The equation you probably had to solve
was $35 - n = 6n$.
This has $n = 5$ as a solution.
Here is the completed puzzle:

	n		**or**		5		
18	$17 - n$	$35 - n$		18	12	30	
	17				17		

E7 If n stands for the top number,
  (a) The middle number is $20 - n$.
  (b) The left-hand number is $18 - (20 - n)$.
  (c) This equation shows that the answer
      to (b) added to n is 24.
$$18 - (20 - n) + n = 24$$
  (d) The solution to the equation in (c) is
      $n = 13$.

E8 Here is the solution to the puzzle.
For this one the top number and the left-
hand number **sum** to 30.

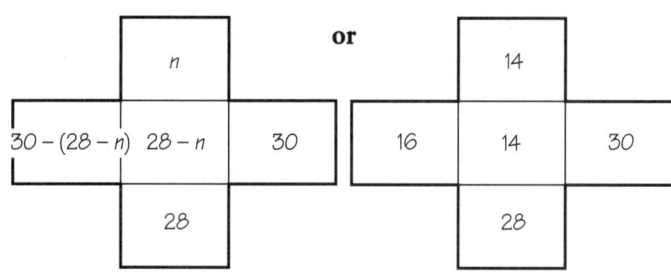

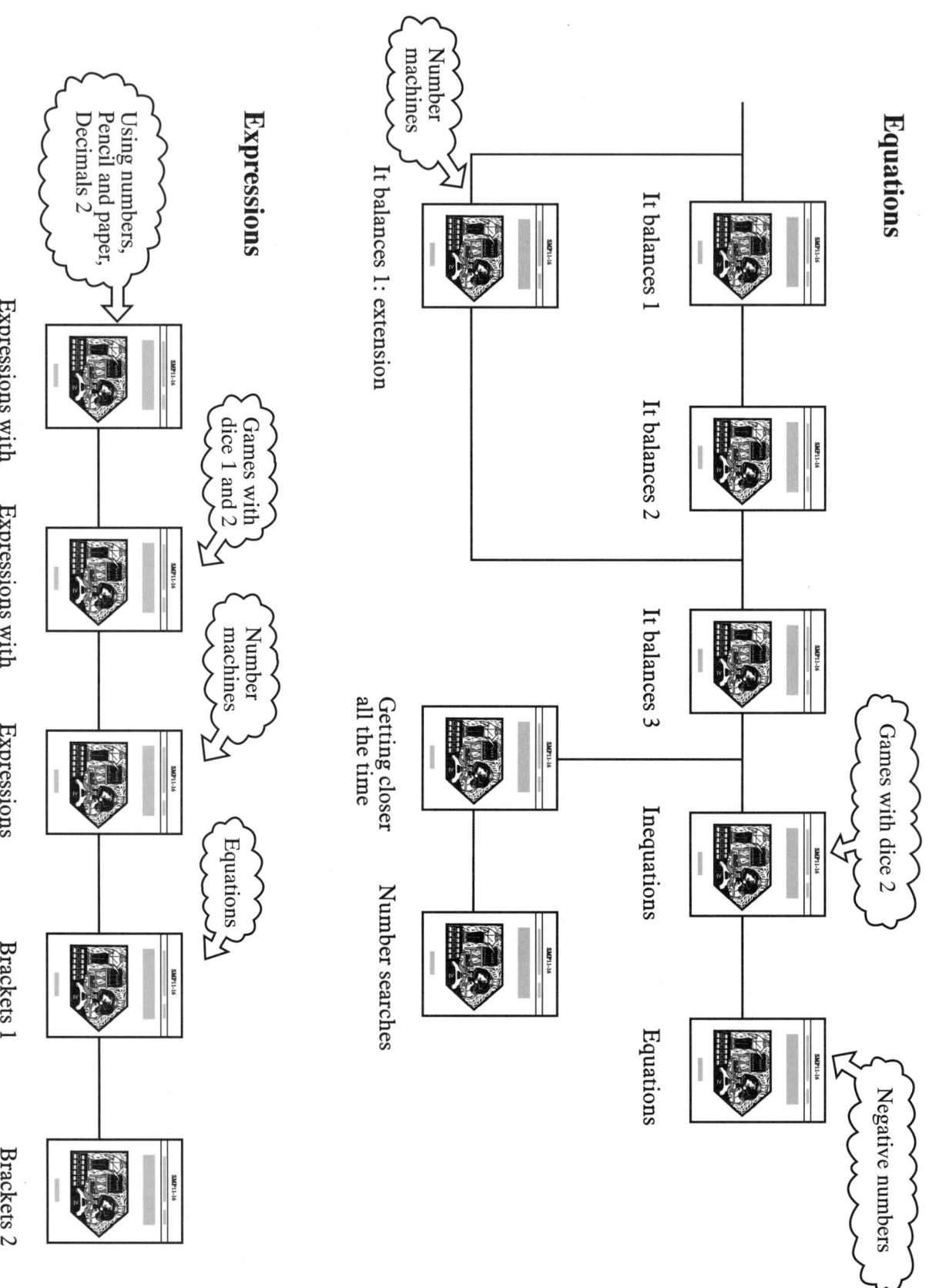

Equations

Number machines

It balances 1: extension

It balances 1

It balances 2

It balances 3

Inequations

Games with dice 2

Negative numbers

Equations

Getting closer all the time

Number searches

Expressions

Using numbers, Pencil and paper, Decimals 2

Expressions with a calculator

Games with dice 1 and 2

Expressions with letters

Number machines

Expressions

Equations

Brackets 1

Brackets 2